D1159633

Animal Habitats

The Deer in the Forest

Text by Linda Gamlin

Photographs by Oxford Scientific Films

Gareth Stevens Publishing
Milwaukee

Contents

Note: The use of a capital letter for a deer's name means that it is a specific *type* (or species) of deer (for example, White-tailed Deer). The use of a lower case, or small, letter means that it is a member of a larger *group* of deer.

A Red Deer stag wades into a stream to feed on the waterside plants, while another watches nearby.

After a snowfall it is much easier to see deer tracks.

Where deer live

Some of the most beautiful animals in the forest are the deer, with their large, gentle eyes, long ears, and glossy red-brown coats. To come across them feeding beneath the trees on a winter morning, their breath turning to steam in the frosty air, is one of the most exciting sights of the countryside. In spring, you may see a deer rearing up on its hind legs to nibble the tender green leaves that are sprouting from the branches. And in summer you might even be lucky enough to find a tiny *fawn*, lying still and silent in the brush, waiting for its mother to return.

There are many different *species* of deer, and they are found in nearly every country of the world. Almost all of them live in forests or woodlands. Some, like the Muntjacs, live in the hot tropical forests of countries like India. Others, such as the Fallow Deer and the White-tailed Deer, live in cooler areas like Europe and North America. Here many of the forests are made up of oak, ash, beech, or maple trees, which lose their leaves in winter. This is called *deciduous* woodland. Other deer, such as the Moose (called Elk in Europe), live in the cold regions of the far north, in forests of pine, spruce, and other *coniferous* trees.

All these different types of forest have one thing in common: They are very good places to hide. Even in deciduous woodland in winter, when the leaves have fallen, there are many quiet, private places. Holly bushes or dense brambles can easily conceal a deer. Thickets of hazel or dogwood can do the same, even though their leaves have gone, because the deer have good *camouflage* in winter. Their brownish-gray coats blend into the background, and a deer's antlers can look just like branches as it stands perfectly still in the middle of a thicket. All deer are very stealthy, cautious animals, well suited to this secretive way of life.

Because they are so quiet and shy, it is often difficult to see more than a glimpse of wild deer. But even though you may not see the deer themselves, they leave many signs of their presence. Their tracks can often be found in muddy areas or along the edges of plowed fields. The prints made by their cloven hooves are easy to recognize, but remember that similar tracks are left by sheep and goats.

A herd of male Fallow Deer, emerging from the forest to feed and looking cautiously around for danger. A magpie has perched on the back of one of them.

Beyond the forest

Very few deer stay in the forest all the time. Often the best food is to be found at the edge of woodland, where trees give way to shrubs. This is where most of the berries grow and the grass is thickest. Farmers' fields also offer many things to eat, especially in winter when other food is scarce. Turnip leaves and other crops will tempt deer out of the forest to feed. They avoid being seen by emerging just after dark or at dawn.

Some deer have ventured even further from the trees, and a few spend their whole lives outside the forest. In Asia, some small deer live in swamps, slipping silently between the tall reeds. The South American *pampas* is home to the Pampas Deer, which can live unseen among its towering grasses. Caribou (also called Reindeer in Europe) have moved out of the northern coniferous forests into the

A herd of Caribou, or Reindeer. These deer have adapted to life in the far north, away from the forest.

bleak open spaces of the *tundra*. Some herds of Caribou go to the tundra just for the summer, while others live there all year round. The tundra is no place to hide, but Caribou can survive there because they are large and live in herds. This gives them some protection against *predators*.

All these types of deer moved out of the forest many thousands of years ago. In more recent times, other deer have been forced out by human activity. Many forests have been felled to make farmland, leaving the deer with nowhere to live. Some of these deer have died out, but others have adapted to the new conditions.

In Scotland, for example, stretches of broad open land called moors are now found where there were pine forests four or five hundred years ago. The Red Deer that once lived in those forests now roam the moors, feeding on grass, rushes, moss, lichen, and heather. They find shelter among tall rocks on the mountain tops, or among the few remaining trees. The moor is not an ideal place for them to live. Almost half the young deer die in the first winter after they are born, victims of cold winds and deep snow drifts. The adults, too, are much smaller than Red Deer living in woodland. But they do survive.

A Red Deer stag. These deer live out on open moors in Scotland and other parts of Europe.

Winter is the hardest time of year for deer, and many die of starvation.

The forest's changing seasons

In coniferous forests, and most types of tropical forests, there are leaves on the trees all year round. But in deciduous forests the trees grow a fresh crop of leaves every spring and lose them again before winter. This cycle of leaf growth and leaf fall is very important—and not just to the trees. It affects all the other plants and animals in the forest, including deer.

The plants growing on the forest floor (often called "undergrowth") are affected because they need light. In summer, the tree leaves cast a heavy shade, so the ground plants must make the most of times when the trees are bare. Their best chance is in the spring, when the weather is warming up but the tree leaves are only just beginning to unfold. Undergrowth plants, such as grasses, ferns, brambles, and hazel, grow quickly then, while the sun is still on them. Later in the year, the growth of these plants slows down, and some die back completely until next year.

In spring and early summer, the young leaves of the trees are tender and juicy, so the deer have plenty to eat.

All these seasonal changes are important to deer. What they can find to eat changes month by month. In spring and summer there are fresh young tree leaves, new grass, and a variety of other foods, so they eat well. Later in the year come the acorns, beechnuts, chestnuts, blackberries, wild rose hips, fungi, and other rich foods. Deer feast themselves on these, and by early winter they are fat and sleek.

Deer need to put on extra layers of fat during the autumn, because in winter they often go hungry. Grass is still available, but it is coarse and stringy now. So the deer must turn to mosses, lichens, and the tough leaves of holly, ivy and other evergreen plants. They may nibble on twigs, or even strip bark from trees to get at the sweet sap in the trunk of a tree. Towards the end of winter, when the new tree buds are beginning to swell, deer can feed on these. If they are lucky, there will be enough food to keep them going until spring, when fresh grass and new leaves appear again.

***Left:* A male White-tailed Deer seen in autumn. The deer feast themselves on nuts and berries at this time of year.**

Being flexible about what they eat allows deer to survive in deciduous woodland. If they needed the same sort of food all year round, they could not cope with the changing seasons.

Food is not the only thing that changes with the seasons. So does the amount of undergrowth. This also affects the deer, especially young ones. Newborn deer are not strong enough to escape from predators. Instead of following their mother, they spend the first few weeks of their lives lying in a well-hidden spot. For them to survive, there must be a dense growth of plants. The undergrowth is at its thickest in early summer, when the plants have completed their burst of spring growth. So it is not surprising that this is when most deer give birth.

In early summer there is thick undergrowth in which fawns, such as this Mule Deer fawn, can hide while their mothers are feeding.

A Mule Deer struggles through heavy snow. This deer lives in the Western states and provinces of the US and Canada, as far north as Alaska. Like other deer, it has a thick winter coat for warmth.

The deer's body

Like most *mammals*, deer have a coat of fur to keep them warm. This is reddish-brown in most species of deer, but it becomes a more grayish-brown in winter. The fur does not actually change color, of course. What happens is that the hairs of the summer coat gradually fall out, or *molt*, as winter approaches, and a new coat grows. This coat is not just a different color. It is also thicker and warmer to keep out the cold winter weather.

Each hair in a deer's coat is hollow, like a tiny drinking straw, but closed at both ends. The space inside the hair is full of air. If you add up all the thousands of little tubes of air in the deer's coat, it comes to a lot of air—almost as though the deer were wearing a life jacket full of air. And when deer go into water, it is exactly as if they *are* wearing a life jacket. The air trapped in the hairs helps them float in water, so they can swim quickly and with very little effort.

A herd of Caribou swimming across a river in Alaska. The air trapped in the hairs of their coat helps keep them afloat.

Left: ***Although their ancestors had five toes as we do, deer now have only two main toes, with two much smaller ones above them.*** **Right:** ***The print left by a deer's cloven, or split, hoof.***

On land, deer are even faster. They have long legs, so they can run swiftly when danger threatens and jump over anything that is in their way. A large deer can easily jump a fence 8 1/2 ft (2.6 m) high. Instead of toes they have hard, springy hooves. These also give a deer extra speed, by pushing firmly against the ground as it bounds away.

Being able to run fast is not enough, however, because predators can move quickly, too. Deer must always be on the alert and notice predators before they get close. So their senses are very sharp, to warn them of approaching danger. Their large eyes are set on the sides of the head, to give them a good view all around. Hearing is also important, and their ears can detect the slightest sound. But the most important sense is that of smell. A deer's nose is so sensitive that it can probably use it in the same way that we use our eyes. In other words, it can build up a "picture of smells," just as we build up a visual picture of the world.

Because their sense of smell is so good, you should always approach deer "downwind" (with the wind blowing into your face) if you want to try watching them. The wind then blows your smell away from the deer, so there is less chance of your being noticed.

Their long legs make deer fast runners, as this Wapiti shows. The name "Wapiti" comes from a Shawnee word meaning "white rump."

A male White-tailed Deer rubbing scent from a gland at the base of his left antler onto a stick.

Smell is important to deer in another way, too. They each have their own personal smell, which is like a signature or a fingerprint. No other deer has exactly the same smell. Deer use these personal smells to mark their *territory*. In this way, other deer will know which part of the forest is already occupied and which is free, just by sniffing the air.

These important personal smells are produced by scent *glands* on various parts of the deer's body. Most types of deer have scent glands just below each eye, at the base of the antlers, and on each back leg. There are also glands on each of the four hooves. The sticky, strong-smelling liquid that oozes out of these glands is usually rubbed off onto twigs and tree trunks. The scent from the foot gland comes off on the ground, especially when the deer scrapes away plants with its hooves to leave patches of bare soil. These are known as "scrapes," and if you see one you will know you are standing in a deer's territory.

A patch of bare earth, known as a "scrape," marks the edge of a deer's territory.

A male Caribou with a magnificent set of antlers. The antlers get larger every year, so only mature males have large ones such as this.

Antlers

Deer are the only animals in the world with antlers. These grow from the top of the male deer's head, between the ears, and are made of bone. They can be very small and straight, as in the Muntjacs of Asia, or huge and branched as in Caribou (Reindeer), Fallow Deer, and Red Deer. The strangest thing about these antlers is that they are dropped, or "cast," every year. Soon after they have been cast, a new pair begins to grow in exactly the same place. Antlers are different from the horns of cows, goats, or antelopes. These are made of horn—like your fingernails—not bone, they are never branched, and they are not cast each year.

After about four months' growth, the deer's new antlers have reached full size. At this stage they are still slightly soft and covered by a thick layer of furry skin, known as "velvet." This contains many blood vessels, which supply the antlers with food while they are growing. Once the antlers are fully grown, the deer gets rid of his velvet by rubbing it off on trees or bushes. After this, the antlers harden and cannot grow any more.

The Muntjacs are Asian deer. They have tiny backward-pointing antlers.

A young Red Deer stag with half-gr antlers which are covered in a laye furry skin, called "velvet."

It takes a lot of energy and *nutrients* for a deer to grow antlers, so throwing the old set away every year and growing a new set might seem wasteful. But it does have advantages. First, it allows the antlers to get larger and more branched every year, as the deer gets older and stronger. Second, if an antler gets damaged, the damage is not permanent because it is automatically replaced.

Having a perfect set of antlers—and as big a pair as possible—is very important to a male deer. The antlers help him impress other males and fight with them if necessary. This is how the males compete for the females in breeding season. The bigger and better his antlers, the more successful a male will be.

A male and female Red Deer, feeding together. The male, on the right, has just cast his antlers.

The cast antler of a Wapiti. All de regrow their antlers every year.

As you might expect, only the males have antlers, because only they need them. This is true for all types of deer except one—the Caribou, or Reindeer. Female Caribou grow antlers every year, just like the males, although they are not as large. No one is exactly sure why they do this, but antlers may help the females defend their young ones from predators.

There is also one kind of deer that has no antlers at all. This is the small Chinese Water Deer, which lives beside rivers and in marshes. Instead of antlers, the male has two long teeth that hang down from its upper jaw, looking a little like Count Dracula's fangs. These teeth, or "tusks," are used to fight rival males, just as the antlers are in other deer.

An Alaskan Caribou, with his fully grown antlers. He has started to rub off the velvet, which is hanging in shreds.

A Wapiti feeding on grass. This deer is sometimes called the "American Elk." But this is not a good name for it, because the European Elk is a completely different type of deer.

Food and digestion

Deer eat many kinds of food, including nuts, fruit, mushrooms and toadstools, lichens, twigs, and even bark. But it is leaves, grass, buds, or crops such as fruit, vegetables, or grains, that make up most of their diet, especially in summer. Some types of deer prefer grass to tree leaves, and they eat much more grass than anything else. These deer are called *grazers*. Other deer eat lots of tree leaves and only a little grass. They are called *browsers*. The Fallow Deer and Wapiti are good examples of grazers, and the Roe Deer and White-tailed Deer are typical browsers.

A Mule Deer stretches up to feed on the needles of a pine tree. Deer can eat many things that do not seem digestible to us.

A Moose feeding on water plants. This deer also lives in Europe, where it is known as an Elk.

Sometimes deer go to great lengths to get their food. When feeding on trees, most deer will stand on their hind legs to reach higher, and the males can use their antlers to pull branches down to them. Roe Deer use their hooves to dig up juicy roots. Moose (Elk) feed on water lily leaves and roots in summer. They wade into lakes and go right underwater to get at them.

But no matter how much food an animal gets, that food is of no use unless it can be digested. This is the process that happens in the stomach and gut, when the food is broken down. Food that cannot be broken down will not be absorbed into the body.

Like other mammals, deer face a problem when they eat foods such as leaves and grass. These are mostly made up of a substance called cellulose which mammals cannot break down. This is why salads don't make you fat—because you can't *digest* the cellulose in them. So how does a deer survive on leaves and grass?

The answer is that deer have some help from *bacteria* living in their stomach. These bacteria *can* break down cellulose, and they go to work every time the deer swallows a mouthful of food. Once the bacteria have broken down the cellulose the deer can absorb it, so grass and leaves are as fattening for deer as bread or potatoes are for us.

The droppings of deer are dry pellets, somewhat like those of rabbits, but larger.

A herd of Sika Deer. These deer came originally from China but now live wild in parts of Europe.

Life with the herd

Some species of deer live together in large groups, known as herds. They include the Fallow Deer, Red Deer, Wapiti, and Caribou (Reindeer). Staying together helps the deer defend themselves against predators, because there is always "safety in numbers." Herds are especially important for females and their young, so they stay together for most of the year. Because of their size and strength, the males have less to fear from predators, and they often live alone for part of the year.

In winter there are other reasons for herding. The normal herds may join together to make much larger groups, containing hundreds, or sometimes thousands, of deer. These deer may be drawn together by a good supply of food when there is little to eat elsewhere. Or they may form these large herds if there is a very deep layer of snow, because the pounding of hundreds of hooves melts the snow and keeps the ground clear. The herd makes a snow-free space for itself, where all the deer can move around freely and feed on the grass. Here in North America, these snow-free areas made by the deer are called "yards."

Even deer that usually live alone, such as Roe Deer and White-tailed Deer, may get together in winter. The groups they form usually number about 20-30 deer, and they split up as soon as spring arrives. These deer do not normally form herds, because they are browsers, living in the forest itself, or in thick vegetation. Here they are well hidden from predators. The deer that form herds are the ones that often feed out in the open, on grass or crops. For them, predators are more of a threat.

Right: Mule Deer herded together in a "yard," where the trampling of all their hooves keeps an area clear of snow.

As they rub the velvet from their antlers and mark out their territory, male deer often damage trees.

Fighting for a mate

For most deer, the breeding season, or "*rut*" as it is called, happens in September, October, or November. In late summer, their antlers finish growing and the males then rub off the velvet. They polish the antlers by rubbing them on tree trunks or hitting them hard against bushes. This often does a lot of damage to the trees and bushes, leaving signs that you may notice, such as frayed bark.

These signs are *meant* to be noticed—not by us, but by other deer, especially other males. They are a male's way of showing off his strength and marking out his territory. As the male is polishing his antlers, scent glands at the base of each antler deposit his personal smell to make the message even clearer.

A Wapiti stag making his mating cry. This cry is described as "bugling" because it sounds like someone playing the bugle.

The male also leaves behind scent from other glands, scrapes the soil with his hooves, and makes a lot of noise. Each type of deer has its own typical rutting noise. Red Deer roar; Wapiti make a sound like a bugle; Moose bellow; Fallow Deer grunt; Muntjac bark like dogs; and Sika Deer whistle.

All this noise and smelliness is often enough to keep the other males away. They give up without a fight, leaving the largest and strongest males to mate with the females. But sometimes fights occur, and if this happens a male may kill his rival with a stab wound from his antlers. Sometimes both males die, because their antlers become locked together. The pair are unable to separate, and after several days they both starve to death.

In most species of deer, the strongest males mate with several females, while the young males do not mate at all. Among Red Deer and Wapiti, for example, the older *stags* round up groups of 20-30 females, known as "harems." Young males cannot compete with older males until they are fully grown, so they have no chance to claim a harem until they are six or seven years old. In other species, such as Roe Deer, each male only mates with one, or perhaps two, females. This means that the young males get more chance to mate.

Wapiti stags fighting for the right to mate. As their smaller antlers show, they are younger than the stag to the left, so they have less chance of mating.

Most deer usually have just one fawn, but this Red Deer mother has had twins. The fawns are two or three months old.

Giving birth

Female deer give birth in late spring, when there is plenty of forest undergrowth to hide in. Those that live in herds leave the others and go off alone to give birth. They usually produce a single young one, or fawn, but some types of deer have twins or triplets, and the Chinese Water Deer has up to six young at once.

The newborn fawn is licked dry by its mother and can stand up before it is an hour old. A few days later it can walk and even run. Young Caribou which live out in the open and have nowhere to hide join the herd then. But most young deer are left lying in the undergrowth for the first week or two. The mother goes off to feed alone, coming back two or three times a day to give the fawn some milk.

While it is alone, the fawn keeps very still and quiet. It has almost no body smell at this age, so predators cannot pick up its scent on the breeze. They are unlikely to see it either, because it is quite well camouflaged. The spots on its coat blend well with the spots of sunlight that fall on the forest floor. This spotted coat is molted when the fawn is a few months old, and it is replaced by the adult coat.

A Moose nurses her young. She will go on giving it milk for six months or more.

When the fawn starts to follow its mother it begins feeding on grass and leaves. But it goes on taking milk as well. This continues for a long time—up to nine months in some types of deer.

Most young deer stay with their mothers for at least a year, until the female is about to give birth again. Just before this happens she chases the one-year-old fawn away. If it is a male, it will probably not return. But a year-old female will come back to the mother and her new fawn a few weeks later.

In a New Jersey woodland, the fawn of a White-tailed Deer waits quietly for its mother to return.

In more remote areas, Mule Deer and White-tailed Deer can still fall prey to Mountain Lions.

Predators and other dangers

Deer fawns face many dangers. Their main predators are foxes and dogs, which can easily kill them as they lie alone in the undergrowth. Coyotes can do this, too. The very small fawns of Muntjac and Water Deer can even be killed by crows and stoats. In the wilder parts of the US and Canada, fawns are often eaten by large predators such as wolves, Grizzly Bears, Mountain Lions, and eagles.

These large, fierce predators will also attack adult deer, and they still kill many in places such as Yellowstone National Park. But throughout most of North America, these large predators have died out. They were killed off because they were in danger to people or to farm animals. The same is true in Britain and most of Europe, although there are still some wolves and bears in Sweden, for example.

With all these large predators gone, you might think that deer would now have a very safe life and that there would be more of them than in the past. But, sadly, this is not true, because humans have taken the place of those natural predators. Millions of deer are shot by hunters every year, and others are killed in accidents, as they try to cross our roads. The *pesticides* that we spray onto crops can also kill deer by poisoning them.

Apart from predators, deer face other dangers, especially in winter. There is not much for them to eat then, and snow can cover up what little there is, so deer often starve. When ice covers lakes and ponds, it is also easy for them to drown. They try to walk on the ice, fall through, and get trapped in the water under the ice. Deep snow is dangerous too, because deer sink into it and cannot run away. If this happens, small predators, such as the Bobcat, may be able to kill them.

The Black Bear does not kill deer, but it will feed on any dead ones that it finds.

A herd of Fallow Deer in a park. Fallow Deer are varied in color. Some are almost black and may be pure white, spotted or unspotted.

Deer and humans

There are some caves in France where our early ancestors painted pictures on the walls showing the animals that they hunted. These cave paintings were made 40,000 years ago, and they show that, even then, people were hunting deer. We have been hunting them ever since, first with spears or bows and arrows, and later with guns. Part of the reason for this is that deer are very good to eat. Their meat is known as venison.

To ensure good hunting, people have always taken care of the deer, and they have even brought deer from one part of the world to another. The Fallow Deer, for example, was brought to Britain from Europe, first by the Romans and then by the Normans. Later it was shipped from Britain to North America, and it now lives in US and Canadian woodlands, too. When animals are taken from one place to another like this, they are said to be "introduced" to the new area.

The early British settlers in New Zealand introduced many types of deer, because they found that their new country had no deer of its own. New Zealand now has Red Deer, Fallow Deer, and Roe Deer from Europe, plus White-tailed Deer, Wapiti, and Moose from North America. There were no deer for hunting in Australia either, so Red Deer, Fallow Deer, and Roe Deer were introduced.

Garden plants can be very tempting to deer, such as this Roe Deer. Unfortunately, they don't like weeds nearly as much!

Sometimes deer are introduced not for hunting, but just so people can enjoy watching them in parks. These deer can escape from the park, and this is how Sika Deer, Muntjac, and Chinese Water Deer spread to Britain. All these deer now live happily in the wild, thousands of miles from their original homes.

Not eveyone likes deer, however. To farmers and foresters they often mean trouble. They may eat crops or nibble the buds and bark from young trees. If there are deer around, newly planted trees need special protection. You may see fences built around these trees, or spirals of hard plastic wrapped around the trunk to keep the deer off.

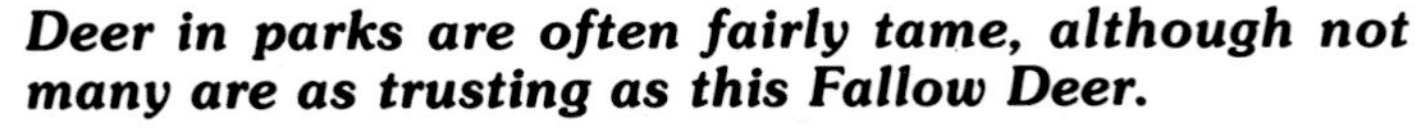

Deer in parks are often fairly tame, although not many are as trusting as this Fallow Deer.

Red Deer in the early morning mist.

The forest by night and day

Deer usually feed after dark, because then they are safer from predators. The best time for them is in the evening or the early morning, just before dawn. At these times it is dark enough for them to feel safe, and yet there is a glimmer of light in the sky, which probably helps them find their food. Only if the moon is out will they go on feeding all through the night.

Deer are not the only animals in the forest to feed mainly at dusk and dawn. Foxes are most active at these times, too, and so are raccoons. Sometimes these animals are searching for the same sort of food as deer, especially berries and other fruit. But they mostly feed on live *prey* such as mice, beetles, and earthworms.

Night is also a busy time for the many small animals of the forest, such as shrews, mice, and voles. Shrews are tiny predators that eat insects and earthworms, but mice and voles are mainly plant-eaters, like deer. Mice like nuts, and they eat a lot of beechnuts and acorns, as deer do. They can also gnaw their way through the thick shells of hazelnuts, which deer would find impossible.

These small animals are preyed on by owls, weasels, and other predators that can hunt in the dark. Bats are around, too, flying along the edges of the woods in search of moths and other insects.

As the sun rises, most of these animals quietly disappear from sight, and the birds' dawn chorus begins. In the daytime, the birds take over the woods. There are far more of them around than at night, and not so many mammals.

Foxes may kill young fawns, but they are no danger to adult deer.

The only mammals that really like the daylight are the squirrels. They eat lots of different food. But beechnuts, acorns, and other nuts are important to them, as they are to some of the birds, including the jays, finches, and nuthatches. In fact, the list of animals that share these nuts with the deer is a long one. Nuts are one of the most important sources of food in the forest.

Luckily for the deer, the other things they eat are not so popular. As we have seen, leaves and grass are difficult to digest. Deer have found a way to digest them, so they can use a food source that is not available to many other animals. Tree leaves are eaten by caterpillars and some other insects, but there are always enough left to feed the deer. Grass is eaten by voles and rabbits, but again, there is plenty to go around.

Woodmice share the deer's nighttime habitat.

Life in the forest

Deer eat many different kinds of plant food, some from inside the forest and some from outside. At one time, deer would have been eaten by large predators such as bears and wolves. In this way, the deer were part of a food chain, as the diagram below shows. They ate only plants, and they supplied food to the meat-eaters, so they were a vital link in this chain. But now the large predators have disappeared in most places, so the deer are mainly killed by human hunters. If you leave out all the animals that no longer live in your area, you will see how much simpler the food chain is now.

Food chain

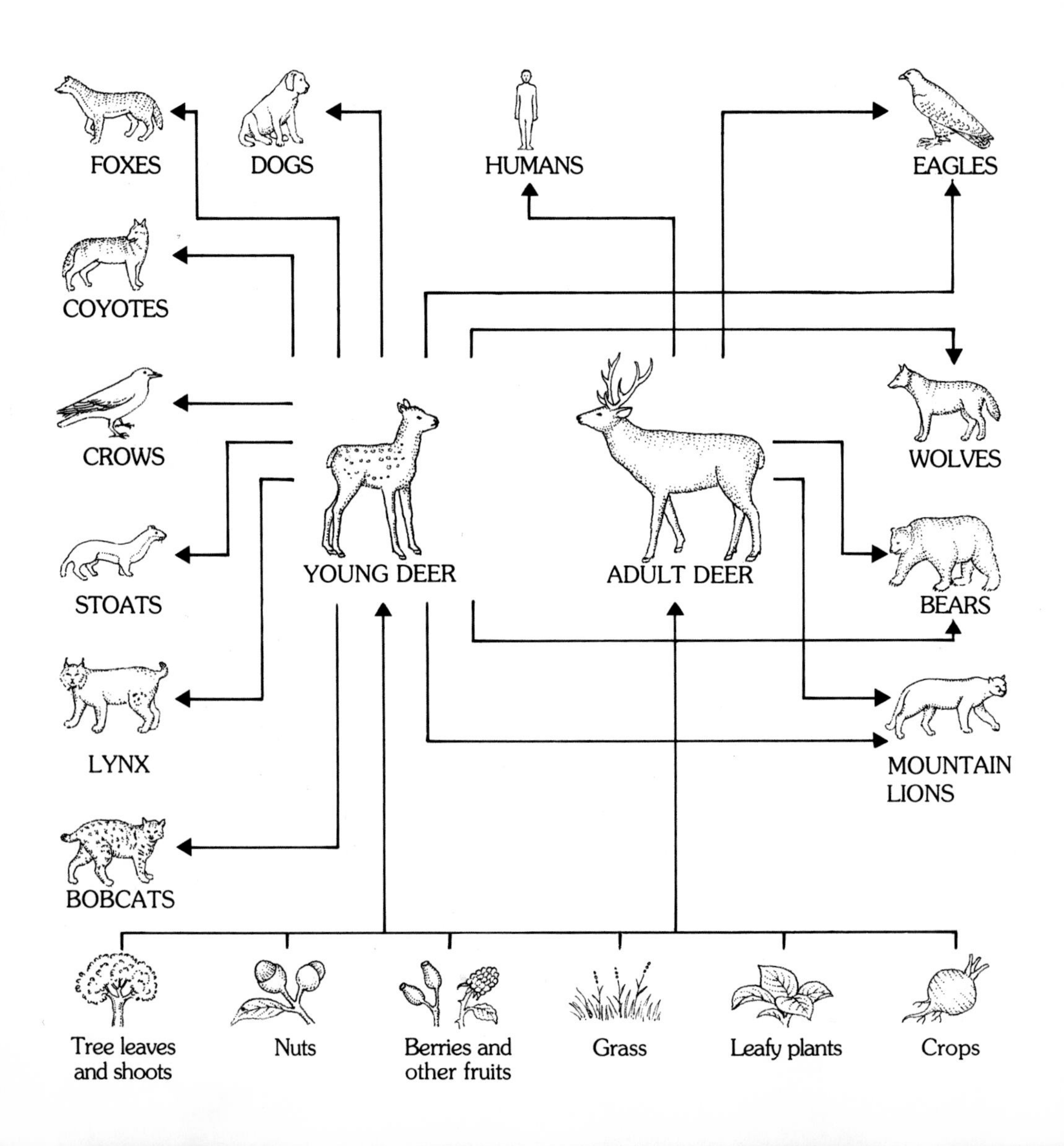

A Red Deer stag with his harem of females during the rut.

Like others in the forest, deer are adaptable animals that can fit in with the changing seasons. They have survived for thousands of years by always being on the lookout for predators and by knowing how to hide or escape. Now they must watch out for the predator with a gun.

This is a far more dangerous enemy than any they have known before. In some places, deer have disappeared completely because of hunting. But where this happens deer are usually introduced again later, using animals from another area. When deer are reintroduced like this, hunting is stopped until their numbers increase again.

People affect the food chain in another way, too. They can make a big difference to what the deer find to eat. Some crops are very useful food for deer, and their numbers may rise if part of the forest is chopped down for farming. The Mule Deer seems to do much better where there is a mixture of forest and fields than it does in pure forest.

But more often the felling of a forest is bad for a deer. They lose the plants that supply them with food as well as the shelter of the trees, and they soon die out. As forests have disappeared, so have the deer. If we want to have deer in the future, we must make sure there are forests and woods for them to live in.

Glossary and Index

These new words about deer appear in the text on the pages shown after each definition. Each new word first appears in the text in *italics*, just as it appears here.

bacteria........very small living things that can only be seen under a microscope. Many bacteria cause diseases, but not all of them do. **17**

browsers.......animals that eat the leaves and young shoots of trees. **16, 18**

camouflage ..colors or patterns that look like their surroundings. They help the animal (or person) wearing them not to be noticed. **3, 22**

coniferous.....trees such as pine, spruce, fir, and larch. They are mostly evergreen and have tough needle-like leaves. Their name comes from the cones which they produce. **3, 4, 6**

deciduous.....(of trees) shedding their leaves at certain seasons of the year, usually in autumn (except in the tropics). **3, 6, 9**

digest ....to break down food into smaller parts, so that it can be absorbed by the body. **17, 29**

fawn............a name for young deer, used especially for the smaller types of deer. **3, 9, 22, 23, 24, 29**

gland...........a small part of the body that produces special liquids, such as sweat, milk, or digestive juices. **12, 20, 21**

grazers.........animals that eat grass and other ground plants. **16**

mammals.....animals with hair or fur that give birth to live young and feed them on milk. **10, 17, 29**

molt.............to shed hair, fur, or feathers and replace it with a new coat. **10, 22**

nutrients ......the important things that are found in food, such as vitamins and protein, which animals need for healthy growth. **14**

pampas........grassland found in South America. Many of the grasses which grow there are very tall. **4**

pesticides.....poisonous chemicals that people spray on crops to protect them from insects and diseases. **25**

predators......animals that kill and eat other animals. **5, 9, 11, 15, 18, 22, 24, 25, 28, 30, 31**

prey.............an animal that is killed by another for food. **24, 28**

rut...............the breeding season for deer. **20, 21**

species ....a type of animal (or plant) which can interbreed successfully with others of its kind, but not with those of a different type. **3, 10, 18, 21**

stag.............a name given to the males of some types of deer, such as the Red Deer. **21**

territory........a piece of land that an animal claims for itself and defends against other animals of its own kind. **12, 20**

tundra ....very cold, wet areas of the far north, where most plants cannot grow. There are only a few dwarf trees and a thin layer of plants, such as mosses and lichens, covering the ground. **4-5**

Reading level analysis: SPACHE 3.1, FRY 4, FLESCH 89 (easy), RAYGOR 4, FOG 5, SMOG 3.5

Library of Congress Cataloging-in-Publication Data Gamlin, Linda. The deer in the forest. (Animal habitats) Includes index. Summary: Text and photographs depict deer feeding, breeding, and defending themselves in their natural habitats. 1. Deer—Juvenile literature. 2. Deer—Habitat —Juvenile literature. 3. Mammals—Habitat—Juvenile literature. [1. Deer] I. Oxford Scientific Films. II. Title. III. Series. QL737.U55G355 1988 599.73'57 87-9916 ISBN 1-55532-298-0 ISBN 1-55532-273-5 (lib. bdg.)

North American edition first published in 1988 by Gareth Stevens, Inc., 7317 West Green Tree Road, Milwaukee, WI 53223, USA

Conceived, designed, and produced by Belitha Press Ltd., London. Printed in the United States of America.
Series Editor: Jennifer Coldrey. US Editor: Mark J. Sachner. Art Director: Treld Bicknell. Design: Naomi Games.
Line Drawings: Lorna Turpin. Scientific Consultants: Gwynne Vevers and David Saintsing.

The publishers wish to thank the following for permission to reproduce copyright photographs: **Oxford Scientific Films Ltd.** for p. 2 (S. Meyers); pp. 3, 4 above, 12 below, 18, 20 top right, 22 above, 31, and back cover (Terry Heathcote); p. 4 below (Richard Barnell); pp. 5, 11 top right, and 17 below (Barrie E. Watts); p. 6 (Raymond Blythe); p. 7 (Sean Morris); pp. 8, 12 above, 13, 24, and front cover (Leonard Lee Rue III/Animals Animals); p. 9 (Stuffer Productions Ltd.); pp. 10 above, 20 above left, and 20 below (Ray Richardson); pp. 10 below, 15, and 16 below (David C. Fritts); pp. 11 top left, 14 top left, top right, and below left (Press-Tige Pictures); pp. 11 below, 16 above, 21, and 25 (Stan Osolinkski); pp. 14 below right and 17 above (Stephen Fuller); p. 19 (Mark Newman); p. 22 below (Harry Engels); p. 23 (Breck P. Kent); p. 26 (David and Sue Cayless); p. 27 below (Alastair Shay); pp. 28 and 29 above (David Cayless); p. 29 below (David Houghton); Bruce Coleman Ltd. for title page (Hans Reinhard).

2 3 4 5 6 7 8 9 93 92 91 90 89